Fire Phone: Out of the Box

Brian Sawyer

Beijing • Cambridge • Farnham • Köln • Sebastopol • Tokyo

Fire Phone: Out of the Box

by Brian Sawyer

Printed in the United States of America.

Published by O'Reilly Media, Inc., 1005 Gravenstein Highway North, Sebastopol, CA 95472.

O'Reilly books may be purchased for educational, business, or sales promotional use. Online editions are also available for most titles (*http://safaribooksonline.com*). For more information, contact our corporate/institutional sales department: 800-998-9938 or *corporate@oreilly.com*.

Editor: Nan Barber
Production Editor: Kristen Brown
Cover Designer: Monica Kamsvaag
Interior Designer: David Futato

July 2014: First Edition

Revision History for the First Edition:

2014-07-25: First release

See *http://oreilly.com/catalog/errata.csp?isbn=9781491911358* for release details.

ISBN: 978-1-491-91135-8

[LSI]

Table of Contents

Preface

After years of speculation and increasingly realistic rumors, Amazon finally satisfied the public's request (or at least curiosity) to add a smartphone to its hardware stable. Like the other devices in that stable (beginning with the original Kindle and its evolutionary iterations, continuing with the Kindle Fire and its later versions, and most recently in the Fire TV), the new Fire phone launches with a number of signature features that make it a uniquely Amazonian product.

The most obvious Amazon-only touch is a new feature that lets you instantly identify and order (via Amazon, of course) just about any product imaginable, from DVDs, CDs, and books (or their electronic equivalents) to soup, nuts, or pretty much anything with a barcode. Though the rationale for Amazon including such a feature is pretty obvious, the innovative implementation (in the form of Firefly, discussed in Chapter 3) is nothing short of magical.

Equally impressive is Dynamic Perspective (see Chapter 2), Amazon's amazing new hyper-3D interface that leaves its defining touches on every screen of the Fire phone's operating system. Add Mayday, the best customer service technology invented since the telephone (see Chapter 4), along with brand-new one-handed navigational gestures (see Chapter 1) and innovative camera features (see Chapter 5), and you have a realistic contender against the iPhone and Android phones that have already established themselves in the market.

Oh yeah, you can also use it to make phone calls.

But you probably already know how to make a phone call from a smartphone, as well as perform all of the other basic functionality shared by the best feature-rich devices, so this concise getting-started guide won't make an effort to cover them. Rather than serve as a definitive manual, this book does not attempt to cover the device's many features in detail. Instead, it assumes some familiarity with smartphones and is meant as an introduction to the new, innovative features available only on the Fire phone.[1]

Prioritizing early information on innovative features over comprehensive coverage of every little detail, this book will get you started *now* on everything that is new and exciting about the Fire phone, right out of the box.

How to Use This Book

This book is organized into the following chapters on the unique features introduced in the new Fire phone:

Chapter 1, Getting Around the Fire
: With the Fire phone, Amazon introduced new navigational tools that change the way you're used to operating a smartphone. This chapter covers the hardware features that make them possible, along with how to use the operating system's three-panel navigation, enhanced Carousel, and brand-new one-handed gestures.

Chapter 2, Dynamic Perspective
: Amazon's new hyper-3D interface influences almost every screen of the new Fire phone. This chapter explains what Dynamic Perspective is, how it works, and how to use it throughout the operating system and within individual apps, such as Maps and games.

Chapter 3, Firefly
: With Firefly, Amazon has completely changed the way you'll identify (and purchase) items in the world around you, with an interface that's as delightful as it is useful. This chapter shows how to use Firefly to recognize (and act on that information) everything from physical media to music or video that is playing around you to numbers, text, and much more.

Chapter 4, Mayday
: While Amazon is already known for their top-notch customer service, Mayday is a game changer, even for them. This chapter

1. Actually, with the exception of Mayday, which is also available on the latest version of the Kindle Fire HD, the features covered in this book are unique to the Fire phone and available on no other *device*, let alone any other *phone*.

shows how to use one of the most effective and useful forms of tech support available anywhere today.

Chapter 5, Camera and Photos
The camera that comes with the Fire phone does everything you'd expect, but it has a few tricks up its sleeve that no other smartphone can claim. This chapter describes how new hardware features, such as optical image stabalization, help you get the best shot. It also shows how to take lenticular photos (as only the Fire can) and use the unlimited photo storage and backup options that Amazon provides for free to every Fire phone customer.

Conventions Used in This Book

Throughout this book, you'll find sentences like this: "Options bar→Menu→Send." That's shorthand for a longer series of instructions that goes something like this: "Tap the middle of the screen to summon the Options bar; on it, tap the Menu button and, from the row that pops up above it, touch Send." Our shorthand system helps keep things snappier than a set of long, drawn-out instructions would.

To call attention to something particularly useful or worth watching out for, the book uses notes like these:

This element signifies a tip, suggestion, or general note.

This element indicates a warning or caution.

Safari® Books Online

Safari Books Online is an on-demand digital library that delivers expert content in both book and video form from the world's leading authors in technology and business.

Technology professionals, software developers, web designers, and business and creative professionals use Safari Books Online as their primary resource for research, problem solving, learning, and certification training.

Safari Books Online offers a range of product mixes and pricing programs for organizations, government agencies, and individuals. Subscribers have access to thousands of books, training videos, and prepublication manuscripts in one fully searchable database from publishers like O'Reilly Media, Prentice Hall Professional, Addison-Wesley Professional, Microsoft Press, Sams, Que, Peachpit Press, Focal Press, Cisco Press, John Wiley & Sons, Syngress, Morgan Kaufmann, IBM Redbooks, Packt, Adobe Press, FT Press, Apress, Manning, New Riders, McGraw-Hill, Jones & Bartlett, Course Technology, and dozens more. For more information about Safari Books Online, please visit us online.

How to Contact Us

Please address comments and questions concerning this book to the publisher:

O'Reilly Media, Inc.
1005 Gravenstein Highway North
Sebastopol, CA 95472
800-998-9938 (in the United States or Canada)
707-829-0515 (international or local)
707-829-0104 (fax)

We have a web page for this book, where we list errata, examples, and any additional information. You can access this page at *http://bit.ly/fire-phone-out-of-box*.

To comment or ask technical questions about this book, send email to *bookquestions@oreilly.com*.

For more information about our books, courses, conferences, and news, see our website at *http://www.oreilly.com*.

Find us on Facebook: *http://facebook.com/oreilly*

Follow us on Twitter: *http://twitter.com/oreillymedia*

Watch us on YouTube: *http://www.youtube.com/oreillymedia*

CHAPTER 1

Getting Around the Fire

While the complete operation of every screen and action available on the Fire phone is beyond the scope of this small getting-started guide to device's unique features, any discussion of the things it does cover wouldn't be worth much if you didn't at least know how to access them first. And, as a significant part of Amazon's innovation is at the heart of navigation itself, it makes sense to start with a brief how-to on finding your way around the device.

Hardware Features

Figure 1-1 identifies the buttons used to access some of the new features we'll cover throughout this book, along with the cameras that enable those features. For now, here's a quick run-down on what these buttons and cameras do:

Power button
: The power button turns the phone on, of course, but it also brings up the lock screen if the phone is asleep and puts the screen to sleep if you press it quickly when it's on (or on the lock screen). Pressing and holding it down brings up a menu of options to power off the phone, restart it, or cancel.

Home button
: If the phone is asleep, pressing the Home button brings up the lock screen. When the phone is in action, the Home button toggles the screen between Carousel view (discussed in "Carousel" on page 6) and a grid view of all the apps on your device (see the screen in Figure 1-1) or stored in the cloud but not downloaded

(toggle between these views by tapping Device or Cloud at the top of the app screen shown in Figure 1-1).

Figure 1-1. Hardware buttons and cameras on the Fire phone

Volume up/down buttons

From the Home screen, tapping these buttons brings up Ringer Volume options (louder, quieter, silent, vibrate, or silence for three hours), as shown in Figure 1-2. Within the Music app or other apps, they bring up Media Volume options, along with controls for advancing tracks, going back, or pausing the song, as shown in Figure 1-3.

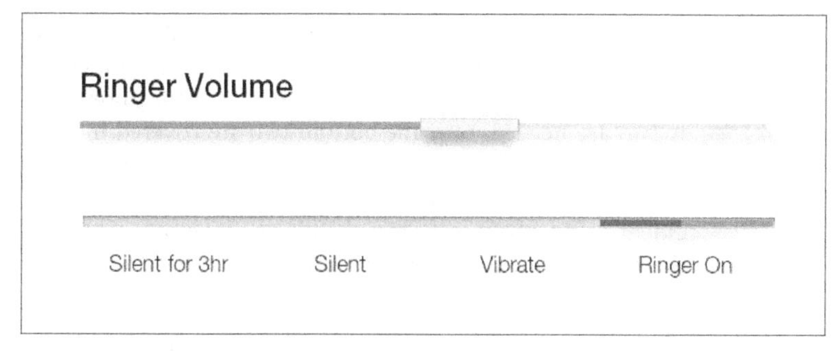

Figure 1-2. Ringer Volume options

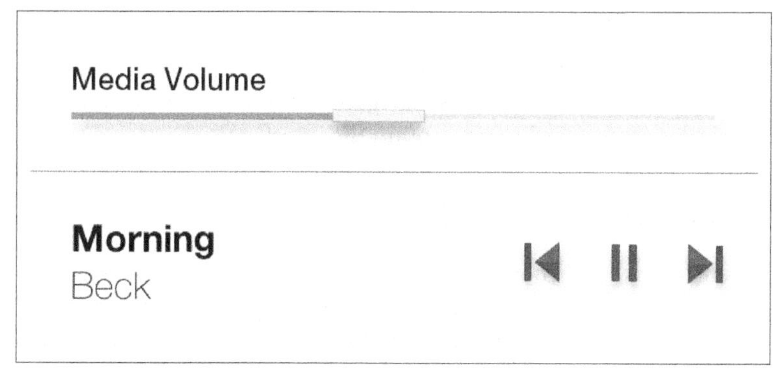

Figure 1-3. Media Volume options

Camera/Firefly button
: Tapping this button and quickly releasing immediately opens the Camera app, whether your phone is already running an app, asleep, or displaying the lock screen. Pressing and holding it (within an app, while asleep, or from the lock screen) opens the Firefly app, which Chapter 3 covers in detail.

Primary front-facing camera
: This is the lens used by the Camera app whenever the front-facing camera is being used.

Dynamic Perspective cameras
: These four cameras work together as sensors to enable all features that rely on Amazon's new Dynamic Perspective technology (covered in detail in Chapter 2).

Generally, Dynamic Perspective depends on at least two of these cameras recognizing the user's face, while the other two account for changes in the way the user holds the phone (only two are needed, so if the bottom two are covered in landscape or portait mode, the other two are sufficient). If you obscure more than two of these cameras (either as a rare accident or on purpose to test them), or move out of view of them, you'll notice the features of Dynamic Perspective fail.

That's about all you need to know about the hardware to get started (I'll assume you can find and use the headphone jack, microphones, and receiver on your own), but the real magic happens when you take a look at the software they open up or make possible.

We'll get to the new features in later chapters, while the rest of this chapter will focus on the new navigation paradigms and associated physical gestures used to get around the interface.

Three-Panel Navigation

If you've used an iPhone or Android phone, you'll feel comfortable with most of the general navigation on the Fire phone (which we won't cover here), but Amazon has introduced a helpful new three-panel system that you might miss if you're just doing what you're used to from your other smartphones.

Accessible from within both the Home screen and the apps written specifically for the Fire, you'll find additional panels on the left and right sides of the screen. Just tap and swipe from the left side of the screen to bring up the left menu and do the same on the right side of the screen for the right menu (or use the one-handed gesture covered in "One-Handed Gestures" on page 8).

From the Home Screen

Figure 1-4 shows what these left and right panels look like from the Home screen.

Figure 1-4. Three-panel navigation on the Home screen

In this case, the left panel for the Home screen gives you quick access to all of your content libraries stored in Amazon's cloud or on your device, including apps, books, photos, and videos. Similar to the Google Now cards on Android phones or the menu you get when you swipe down on from the top of an iPhone Home screen, the right panel brings in pieces of information from your apps that might be immediately useful, such as snippets from new emails, text messages, the weather, appointments in your calendar, etc.

Within an App

Generally, within an app, the left panel provides a quick menu of options and the right panel brings information that might be immediately useful, interesting, or even delightful. As just one example, check out the three panels of the Music app, shown in Figure 1-5.

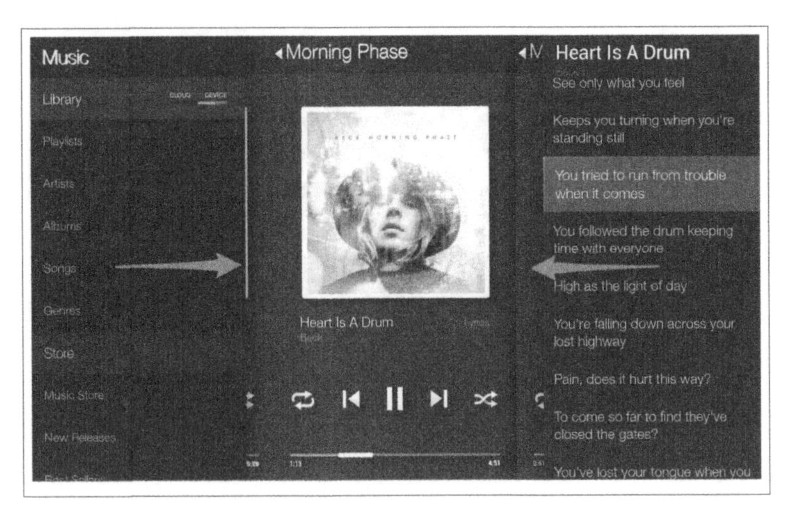

Figure 1-5. Three-panel navigation in the Music app

Here, the left panel gives you access to your playlists, artists, songs, etc., as well as a direct link to Amazon's Music store, while the right panel provides real-time lyrics as the song plays (when available). The highlighted text advances as the song plays, and you can even scroll down further in the lyrics, tap a line, and the song will jump forward to the part where the tapped lyric plays.

This is one of those features that is both useful and delightful. Be sure to test the left and right panes from within all of your apps to see what other things you might otherwise be missing.

Carousel

If you've used any Kindle Fire model, you'll already be familiar with the Carousel, an area on the Home screen that highlights recently used apps, books, or other media in a rotating display you can navigate by tapping and sliding right (for more recently used items) or left (for older items). While the Kindle Fire places the Carousel directly above a grid of your regular content library on the same screen, the Fire phone gives the Carousel its own screen (toggle between the Carousel view and the library view by tapping the Home button, as discussed in "Hardware Features" on page 1).

Why give the Carousel its own screen? Well, beyond just not competing with the rest of your library and cluttering up the smaller real estate offered by a phone, Amazon also beefed up its Carousel for the Fire phone. With the enhanced features it now has, it really needs its own screen.

The newly enhanced Carousel view now includes not only the "hero" icon of the app, book, or item in your history, but also small bits of useful content updated in real time. As you can see in Figure 1-6, the Books app (not a lot changing in real time) shows only recommendations for purchases based on other books you're reading. But Zillow has enhanced their app to include property information about homes for sale or recently sold in your area (you can specify what you're looking for within the app), and USA Today shows headlines for top news stories. Just click on the snippet to open the app and go directly to that item.

Figure 1-6. The enhanced Carousel screen

Each app developed to work with the enhanced Carousel provides different information that is most likely to be useful to the user immediately. For example, the Email app shows incoming messages below the app icon in Carousel view, allowing you to take action on a message without even opening the app (if you want to delete it) or opening the app directly to that message for response.

As handy as it is, the Carousel is limited to recently used items, so older items might quite a bit of scrolling to find. To skip the Carousel and go right to the full app grid, just swipe up from the bottom of the screen (swipe down from the grid screen to get the Carousel) or tap the Home button (tapping the Home button from the grid view opens the Carousel view).

One-Handed Gestures

Beyond their new swiping navigational options introduced in "Three-Panel Navigation" on page 4, Amazon has invented brand-new one-handed gestures (in practice, swiping usually requires one hand to hold the phone and the other to swipe) that you might completely miss if you're not looking for them but which could really help your productivity once you know about them.

Knowing that these gestures are made possible by Amazon's new Dynamic Perspective technology (see Chapter 2) might help you understand how to use them a little better or troubleshoot them when they occasionally require more effort. But for now, you don't really need to know much about what goes on under the hood to use them.

These new navigational tools work without any need to touch the screen. You simply tilt, shift, or swivel in various directions.

Peek

Gently angling the phone to the left or the right, as shown in Figure 1-7, reveals extra, "layered" information about items shown on any given screen.

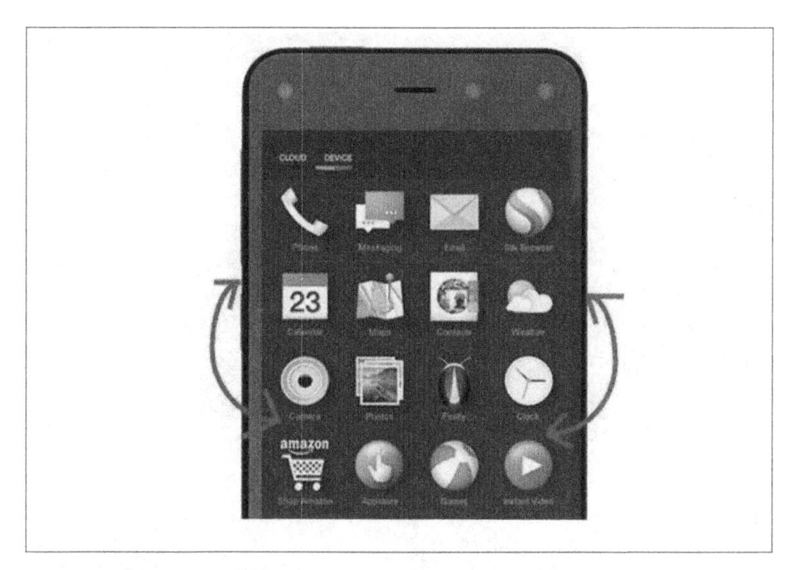

Figure 1-7. Peeking gesture for "layered" information

The extra information changes based on what's currently displayed on the screen. Figure 1-8 shows the extra information that appears when you peek into the left panel on the Home screen.

The menu items appear three-dimentional, because the tilting shows the "sides" of the letters when you angle the phone.

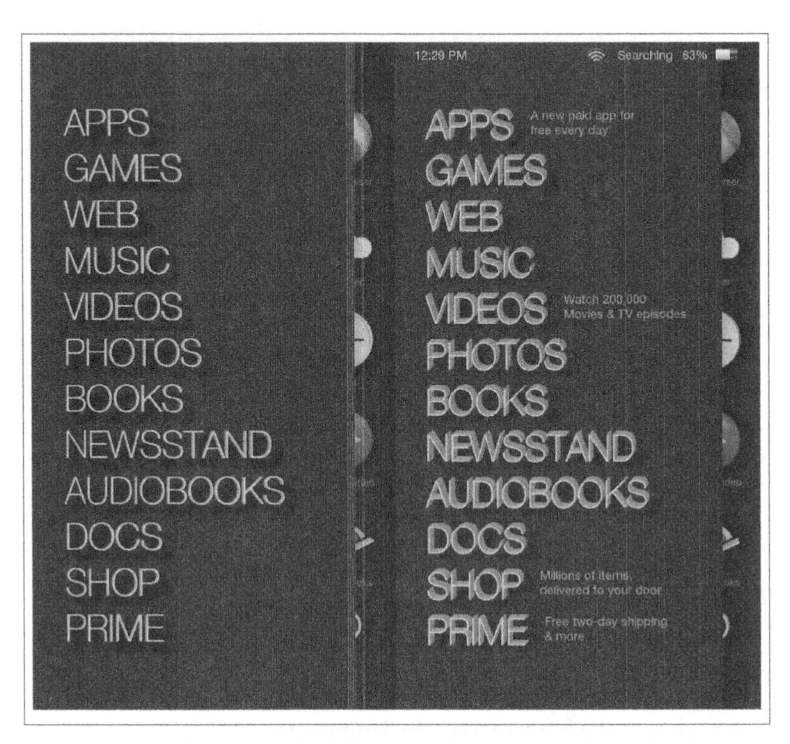

Figure 1-8. Peeking into the left menu of the Home screen

Even if it just presents a side view of the icons displayed, peeking has an effect on almost every screen with Fire phone. Go ahead and try it out wherever you are to see what extra information you might find.

For example, if you've searched for restaurants using the Maps app, peeking reveals Yelp reviews for the locations plotted on the map, as shown in Figure 1-9.

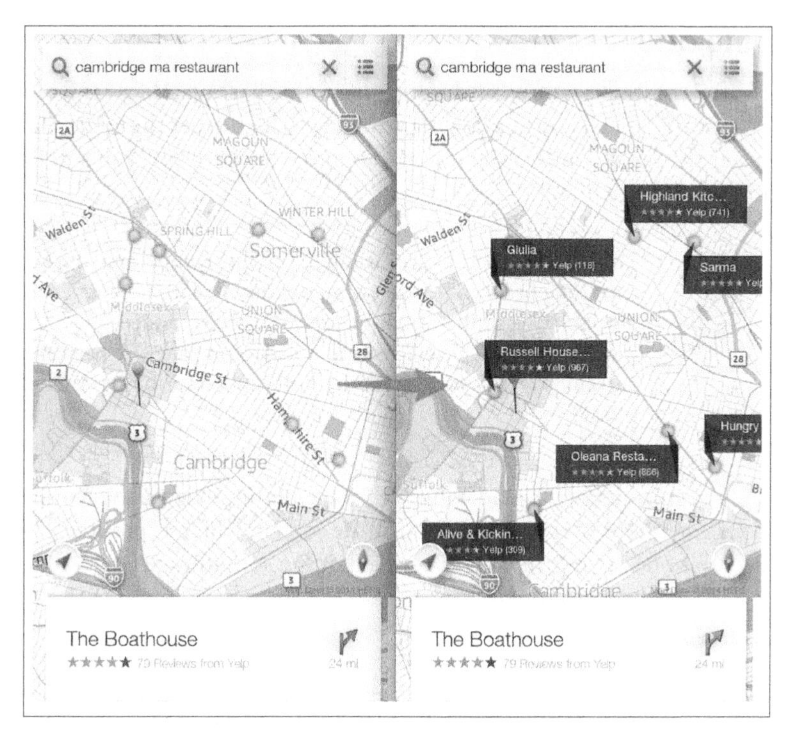

Figure 1-9. Peeking at Yelp reviews in Maps

In addition to adding information, peeking allows you to see under or around icons or titles that overlay and obscure the content (such as the search bar or the My Location icon in the lower-left corner of the Maps screen).

Tilt

Tilting is done using the same angling action as peeking (see Figure 1-7), but more quickly and with more force. Shifting the phone sharply in this way has the same effect as swiping from the edge of the phone, as discussed in "Three-Panel Navigation" on page 4.

Tilting the left side of the phone toward you displays the left panel of the Home page or apps (tilting the left side away from you again sharply dismisses it). Tilting the ride side toward you displays the right panel (and tilting the ride side away from you again dismisses it).

Auto-Scroll

A similar tilting gesture, but angling the phone base up and away from you, automatically scrolls down a web page. The steeper the angle, the faster the scroll. Pulling the top of the phone back toward you scrolls in the opposite direction.

Though auto-scroll was featured in the Kindle app during Amazon's press conference announcing the Fire phone, it was not included in the initial release. When it is added in a later software update, it will feature a lock icon that will keep the current pace of your scrolling without requiring continued tilting.

This gesture lets you read long web pages (and, eventually, Kindle books) without ever having to touch the screen, which is particularly useful if one of your hands is occupied doing something else.

Swivel

Sharply swiveling the top corner of the phone in a tight arc from right to left (as shown in Figure 1-10) or left to right reveals the Quick Actions bar shown in Figure 1-11.

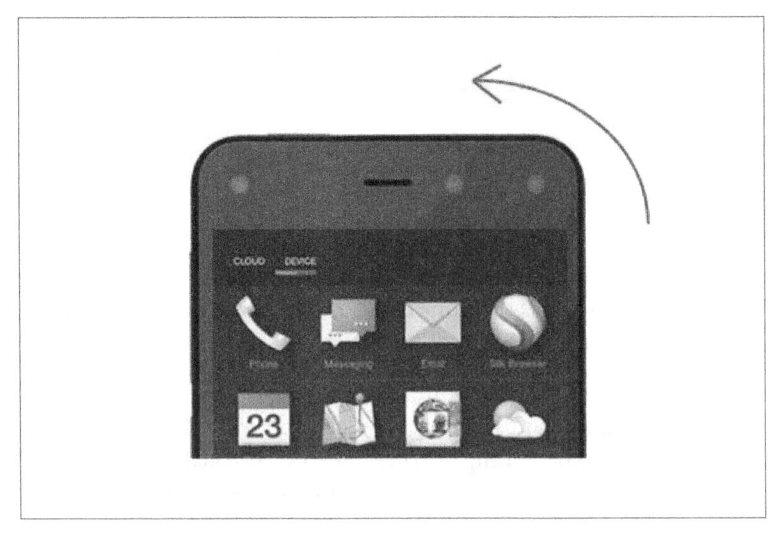

Figure 1-10. The swivel gesture

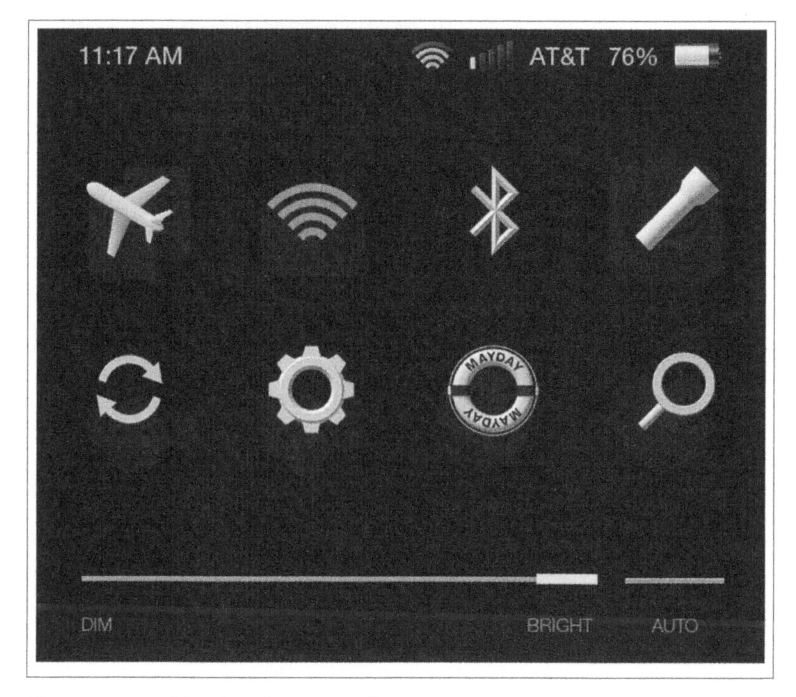

Figure 1-11. The Quick Actions bar

From this menu, you can quickly put your phone in Airplane Mode, connect to a wireless hotspot, turn Bluetooth on or off, access the flashlight, sync your device, access the larger settings menu, search the device, adjust your screen's brightness, or connect with Mayday for customer support (as discussed in Chapter 4).

CHAPTER 2

Dynamic Perspective

You think the Mona Lisa's eyes follow you around the room when you're staring at her? Just wait till you get a load of the lock screens that ship with the Fire phone, which feature a new technology Amazon calls Dynamic Perspective.

In fact, it occurs to me that none of the lock screens that come with the Fire include human faces (no eyes looking out at you), which makes sense, as that would likely seem way too real and push them from the "Wow!" side of the amazement spectrum to the "Creepy!" side.

Dynamic Perspective is a kind of hyper-3D view that allows you to peek around objects in the foreground to see what's behind them, all without requiring special glasses or anything more than your own eyes on your own face. Instead, it manages this feat by using four dedicated cameras on the front of the device (as discussed briefly in "Hardware Features" on page 1), which act as sensors to determine the user's position relative to the image on the screen.

The effect is a pretty spectacular-looking shifting image, with depth and responsiveness unlike anything seen before on a consumer device.

Lock Screens

Though the potential applications for Dynamic Perspective in the hands of the independent developer community—Amazon has made its software development kit (SDK) and associated application programming interfaces (API) publicly available for that purpose—are exciting to think about, for now, the most visible and publicized implementation of this new technology is found in the lock screens that ship with the Fire phone.

Figure 2-1 shows an example of a few different views of the Easter Island lock screen. What it doesn't show is the many other angles and versions of the image you can see when you actually view it on the Fire screen. Tilting the phone away from you shows much more of the night sky, complete with constellations and shooting stars. That little UFO on the rightmost side of the figure zooms around in the background before settling there. And notice how the date moves behind the foremost statue in the image. There's so much going on in each of these images, you really need to view it in person and get lost in it yourself.

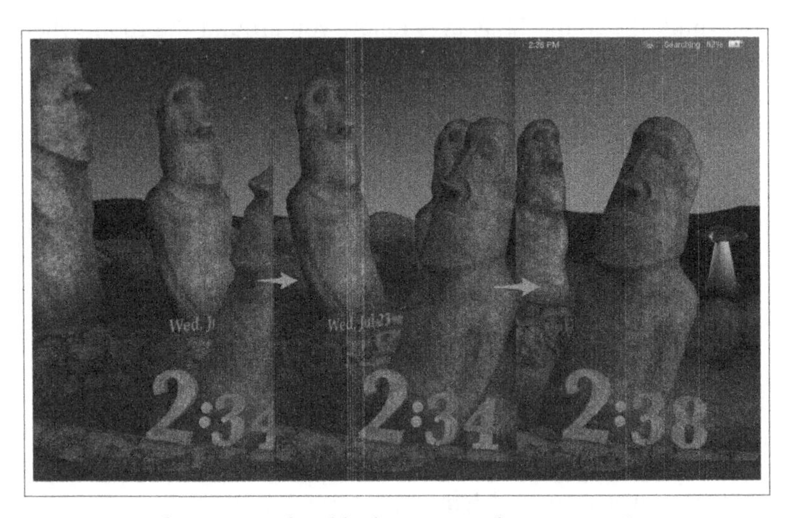

Figure 2-1. The Easter Island lock screen with Dynamic Perspective

If you don't think too much about the process behind it, the effect seems like the result of tilting the device. While that might be the case in practice, what's actually going on is the phone's sensors following your face as your viewing angle changes.

To test the effect (and perhaps be more impressed), start by setting the lock screen with one of the lenticular images that come with the phone. Go to Settings→Lock Screen→ "Select a lock screen scene" and tap on any image to preview it (there are just fewer than 20 mind-blowing options at the time of this writing). When you find one you like, tap the check mark in the lower-right corner to save the image to your lock screen.

Tap the Power button to put the screen to sleep and tap it again to bring up the lock screen. Now, set the phone down on a table and look at it from above, shifting your perspective but leaving the phone still. Watch how the image shifts based on *your view* of the screen, not the position of the phone independent of you.

Maps

Another useful and impressive use of Dynamic Perspective is obvious in the Maps app that Amazon developed specifically for the Fire phone (using Nokia's HERE Maps database). It's similar to Google Earth's 3D view, but with the benefits unique to Dynamic Perspective (tracking based on the user's view, seeing around objects by peeking, etc.).

For example, Figure 2-2 shows three different views of the John Hancock building in Boston, including one (on the right) that allows us to "peek" at its Yelp reviews.

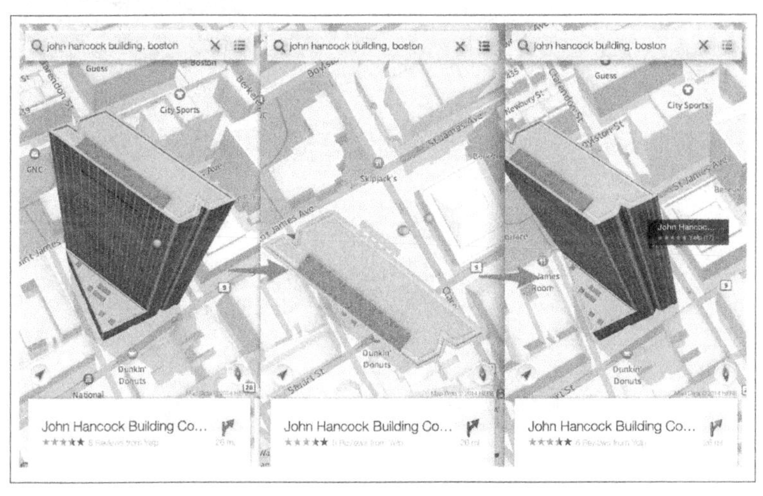

Figure 2-2. The John Hancock building in the Maps app

Tilting the phone (or shifting your position while the phone remains still) allows you to see around the building to find what's behind it, all while retaining a richer 3D view of the map.

Games

Game developers are going to have a lot of fun with Dynamic Perspective, and a few have already gotten started. Already in the App Store, games like To-Fu Fury, Amazing, and Snow Spin are taking immersvie 3D gaming to a new level.

In the first-person adventure game Amazing, you begin in a labyrith and find your way through by physically turning your head to see around walls and down different passageways. Figure 2-3 shows the character turning to his left (as the user turns his head) to see a new path to follow.

Figure 2-3. The Amazing game

Similarly, the snowboarding game Snow Spin (shown in Figure 2-4) allows you to steer your snowboard using your head movement (it also has an option to steer by swiping).

Figure 2-4. Snow Spin

These games are just the beginning in this space, but they show what's possible using Dynamic Perspective as a platform. They're also useful to play around with, not just because they're fun, but also because they help you get used to using this new type of interface. When the new implementations we haven't even thought of yet arrive, you'll be ready.

The Fire Operating System

As mentioned briefly in "One-Handed Gestures" on page 8, though the use of Dynamic Perspective is most obvious in the features and apps that showcase it, it's also the technology under the hood that powers the Fire phone's new navigation options, one-handed gestures, and all the subtler touches that make the Fire operating system unique.

Since we've addressed those other functions elsewhere, this chapter has focussed on the features and apps that specifically call attention to its use, though you should keep in mind that Dynamic Perspective is a foundational technology used throughout the entire operating system.

CHAPTER 3
Firefly

Like Dynamic Perspective (Chapter 2), Firefly is another new technology introduced in the Fire phone that is sometimes hard to distinguish from magic. By pointing the Fire's rear-facing camera at the world around you, the phone can recognize items it identifies in its field of vision, from stuff you can buy (such as DVDs, CDs, and books) to printed information you can act on with other apps in your phone (such as the items discussed in "Phone Numbers, URLs, and Email Addresses" on page 26).

After recognizing an item, Firefly provides contextual details relevant to the particular item you've found (more on these specific details in the sections that follow), allowing you to purchase the item (a benefit for both Amazon and you, as your complimentary Prime membership kicks in) or interact with it in some other way. Firefly does all of this so quickly and seamlessly, with some of the most obscure items you can find, that it certainly seems like magic.

Amazon claims Firefly recognizes more than 100 million items (including 245,000 movies and TV episodes, 160 live TV channels, 35 million songs, and 70 million miscellaneous products), an impressive number that isn't easy to doubt once you start to try stumping it.

To get started, you can just navigate to Firefly by tapping its icon in the Carousel or app grid, but there's a much easier and quicker way. The Camera button (refer back to Figure 1-1) doubles as a Firefly button. Just press and hold to open from anywhere (within an app, from the lock screen, or while the screen is asleep), and you're ready to start scanning.

Books

True to Amazon's roots as a book retailer and developer of hardware for reading ebooks, Firefly excels in recognizing physical publications of the printed word. See Figure 3-1 for an example of it in action.

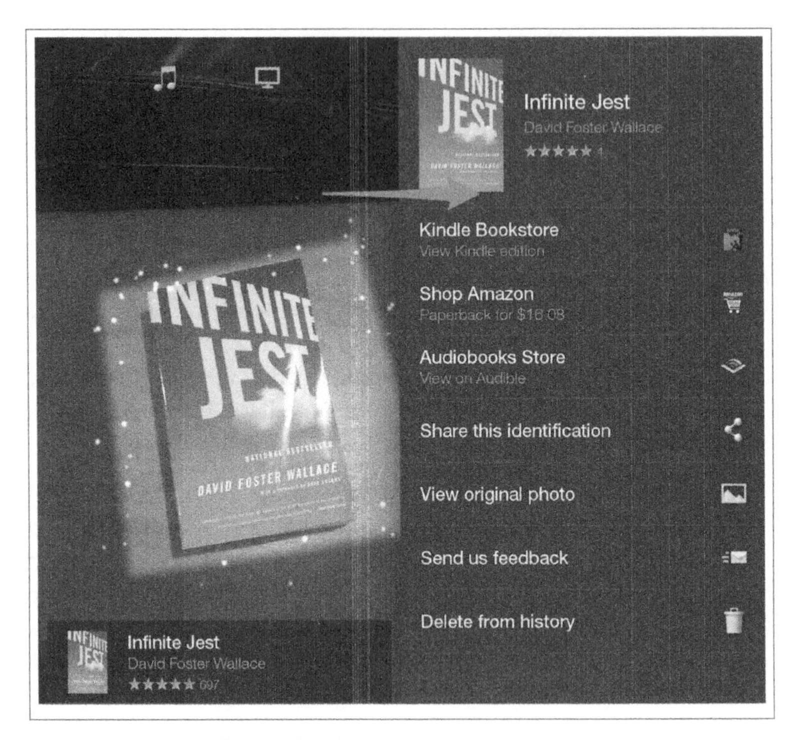

Figure 3-1. Identifying a book

As you can see, when the book is identified (as with any other identified item), the title appears in a tab at the bottom of the screen. Just tap the tab to open the details page shown on the right in Figure 3-1.

In the case of books, you have the option of buying the Kindle version of the title, the paperback (or hardcover, likely prioritized by which version is available or currently selling more copies at Amazon), or the audiobook via Audible (an Amazon company).

Tapping "Share this identification" brings up options to send the link to the identification via Twitter, Facebook, Email, or Messaging. You can also add the link to your Notes app from this page.

Movies and TV Shows

Firefly also recognizes movies and TV shows, both in the way you've come to expect with books and in an even more magical way that might surprise you. First, of course, it recognizes DVDs, as shown in Figure 3-2.

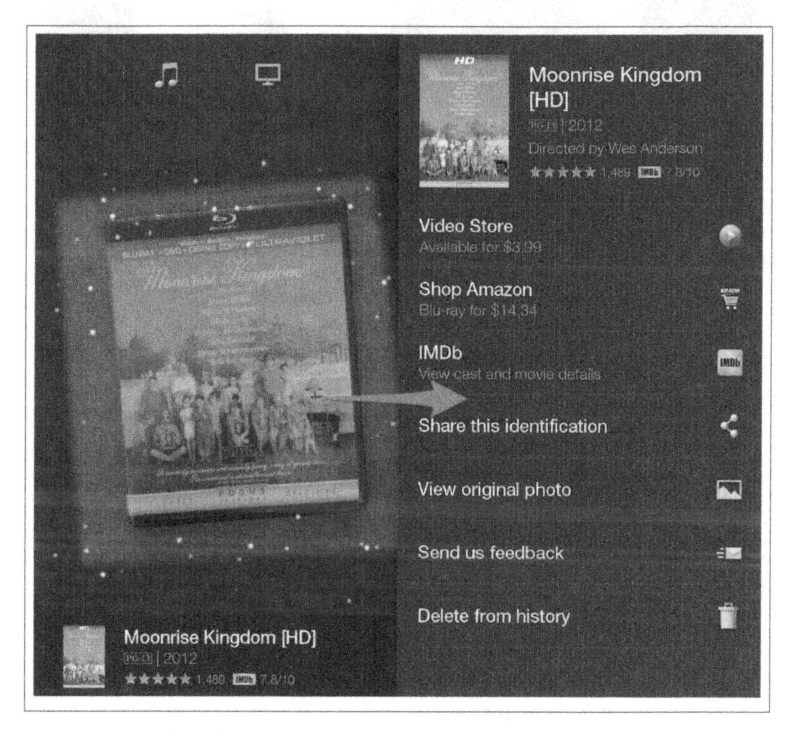

Figure 3-2. Identifying a DVD

From the details screen, you can go straight to the Instant Video store to buy or rent it and start streaming immediately (you can also purchase the DVD). Tapping IMDb app opens a wealth of information from IMDb (the Internet Movie Database, also an Amazon company).

But the more amazing and magical way Firefly can identify movies and TV shows is by *actually watching them*. To recognize a movie or

TV show you're watching, open Firefly, point it at your TV screen (or whichever screen is showing the content), and tap the TV icon highlighted in Figure 3-3.

Figure 3-3. Identifying a TV show

As you can see in Figure 3-4, Firefly recognized *Sherlock*, Season 2, Episode 3 (titled "The Reichenbach Fall").

In addition to the options you get for a DVD, if it's a show Amazon offers in its Instant Video library, the details screen tells you which actors are in the current scene you're viewing (see the In Scene section of Figure 3-4) and how far into the episode you are (right next to the In Scene heading). In this case, I was 3:01 into the episode when I identified it using Firefly.

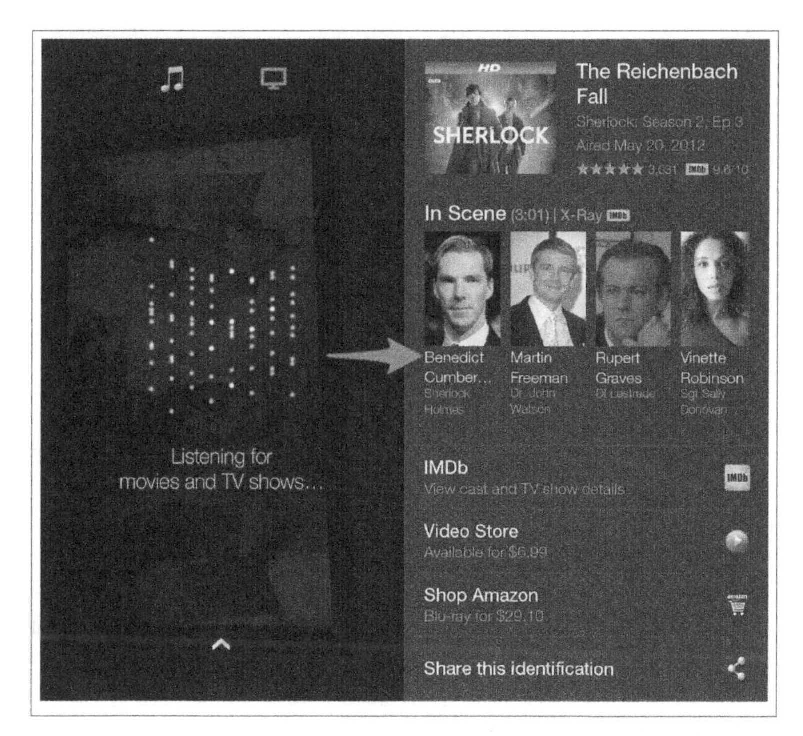

Figure 3-4. TV show details and options

Music

Identifying music works in the same way as movies and television, both for physical media and for songs that are playing near your phone, but it also offers a few additional features specific to music.

So, recognizing a CD is pretty much the same as recognizing a DVD. But, just as Firefly can *watch* a movie or TV show, it can also *listen* to a song. Just open Firefly and tap the musical-note icon (shown next to the TV icon in Figure 3-3).

Figure 3-5 shows the identification screen you'll see, along with the options to use the identification with the StubHub and iHeartRadio apps.

To see the options for StubHub and iHeartRadio, you must first have accounts with those services and install their respective apps (both free) from the App Store.

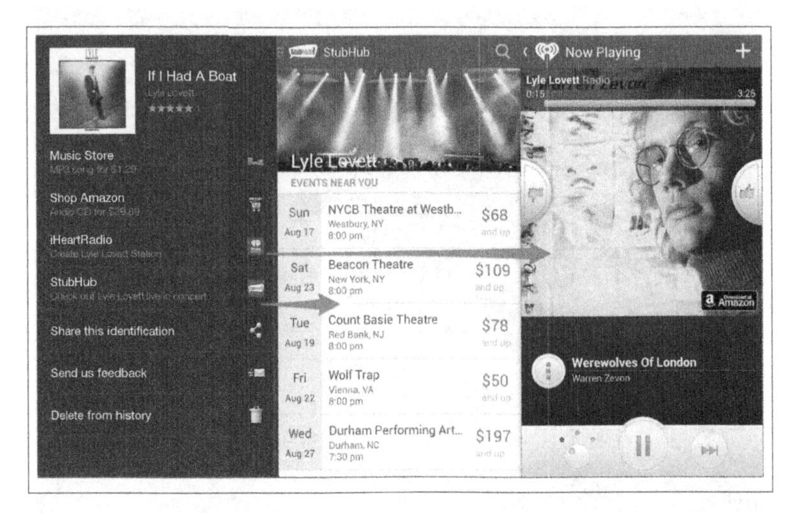

Figure 3-5. Opening StubHub and iHeartRadio apps from Firefly

As you can see, when available, StubHub shows the artist's concert touring schedule (tap a date/venue to purchase tickets), and iHeartRadio creates a station (a playlist of similar artists and songs) from your identified song.

Phone Numbers, URLs, and Email Addresses

While Firefly is great for recognizing *products* and giving you an easy route to purchasing them, that's not all it can do. It can also recognize numerical and textual information that you can immediately act upon from other apps on your phone. Sepecifically, it recognizes phone numbers, URLs, and email addresses on printed material.

Figure 3-6 shows a phone number and URL identified on a member newsletter for Boston's Museum of Science.

Figure 3-6. Identifying a phone number and URL

Make sure you look at *everything* Firefly identifies, not just what you're looking for, and you might be surprised by what it finds. I didn't even notice the book cover for *Salt Sugar Fat* in the blurb further down the page, but Firefly obviously did.

To take action on these items, just tap the phone number to place a call or tap the URL to open it in the Silk browser. Tapping a recognized email address opens the Email app to compose a message.

Everything Else

If you use Firefly enough, you'll continuously be amazed by what it identifies, so no list will be complete, especially as Amazon continues to add features in later software updates. But even now, beyond product images and text/numbers, you can also scan barcodes and QR codes for whatever Firefly's image recognition might miss.

Beyond that, some identifications lead to relevant products that might not be quite exactly what Firefly has recognized but are likely still interesting or useful to you. For example, identifying a playbill for a community theater musical production brings up options to buy the cast recording for the Broadway show, along with the iHeartRadio options discussed in "Music" on page 25, as shown in Figure 3-7.

Figure 3-7. Cast recording details from a musical playbill

I once received a print of an antique packet label for sunflower seeds, which now hangs in my kitchen. On a lark, I thought I'd see what Firefly made of it, as shown in Figure 3-8.

Figure 3-8. A print sold on a shopping bag

Now I know that Amazon sells a shopping bag with the same print.

History

Now that you've scanned everything around you, you might want to revisit some of those identified items. To review your history (shown in Figure 3-9), just swipe up from the bottom of the screen from within the Firefly app.

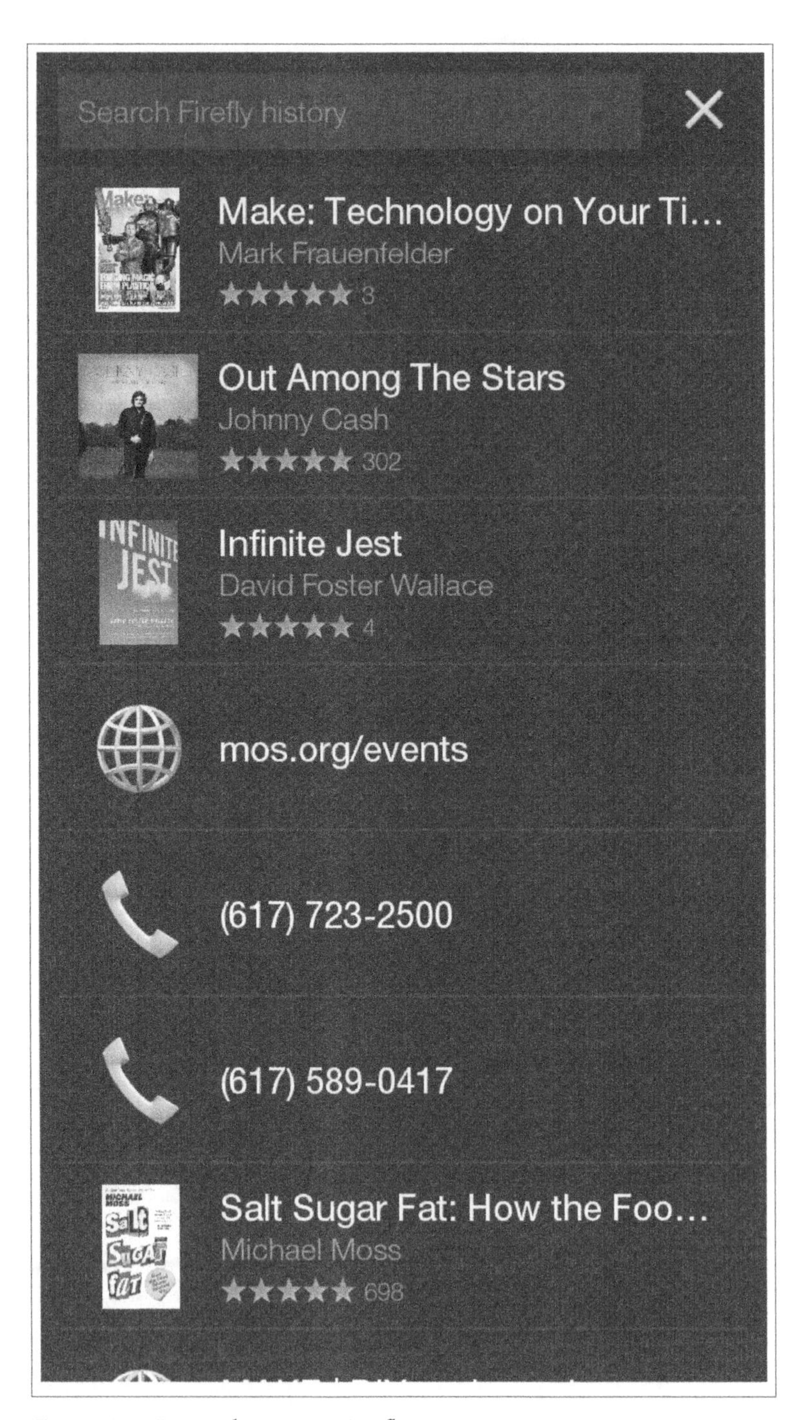

Figure 3-9. Recent history in Firefly

Recent identifications also appear below the Firefly hero icon in the Carousel, as shown in Figure 3-10.

Figure 3-10. Firefly icon in the Carousel, with recently identified items

Just tap the item to open its details page in Firefly.

The Future

As amazing as Firefly is now, it's only going to get better. Amazon has already announced that text *translation* (scan printed text in one language and translate it into another) is on the roadmap for future software updates, and other innovative uses will surely be introduced as they continue to enhance this already incredibly useful feature.

Keep scanning today, and who knows what you'll end up recognizing tomorrow.

CHAPTER 4
Mayday

Amazon prides itself on its customer service, and its mission to put the customer's needs first is at the heart of everything it does, so it's not surprising that they're innovating in this department as well. Much more than just a gimmick, their new Mayday feature is one of the most effective and useful forms of tech support available anywhere today.

Though innovative and popular, the Mayday feature is not unique to the Fire phone. It was introduced in the most recent versions of the Kindle Fire HD and Kindle Fire HDX (all sizes), which are the only other devices that offer it.

Mayday promises one-on-one guidance from a knowledgeable expert (Amazon calls them Tech Advisors) about whatever trouble you might be having with your device—from setup questions to troubleshooting to assistance with features—within about 15 seconds of placing your call.

To get started, just bring up the Quick Actions bar (see "One-Handed Gestures" on page 8) and tap the Mayday icon, highlighted in Figure 4-1.

Figure 4-1. The Mayday icon in the Quick Actions bar

Before long (Amazon says the wait is usually less than 15 seconds, which has been the case every time I've used it), an Amazon Tech Advisor will appear on your screen, as shown in Figure 4-2.

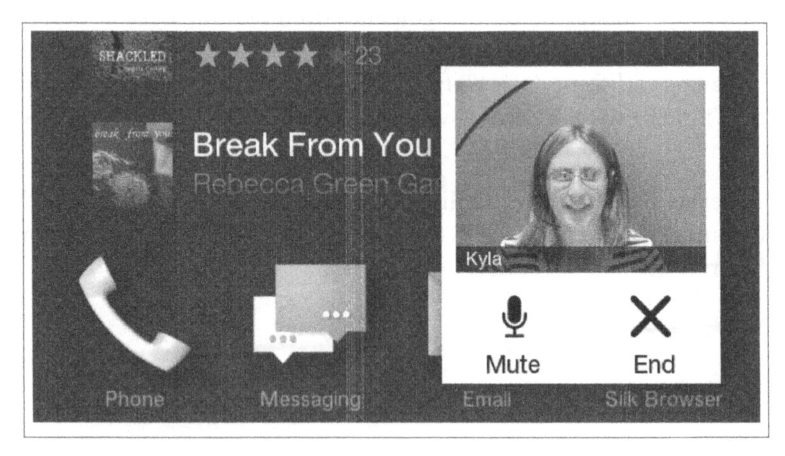

Figure 4-2. Starting a Mayday call with an Amazon Tech Advisor

Once you're connected the Tech Advisor will be able to help you in a variety of ways.

Video Chat

The first and most obvious way the Tech Advisor can help is just by talking to you, listening to your problem and offering verbal solutions, via a one-way video chat (you can see the Tech Advisor, but she cannot see you).

Though it requires the least technology, this somewhat old-fashioned method of tech support (compared with the methods described in the following sections) will likely be enough to handle most of the problems you experience. The Tech Advisor can see your screen, ask questions, and often comment on what you might be doing wrong, without having to take any additional action of her own.

Drawing on Screen

If talking it out doesn't solve your problem and you need a little more hands-on assistance, there's much more a Tech Advisor can do to help, including drawing on your screen to help you find what you're looking for.

In Figure 4-3, Tech Advisor Kyla helpfully illustrates the location of the Home button.

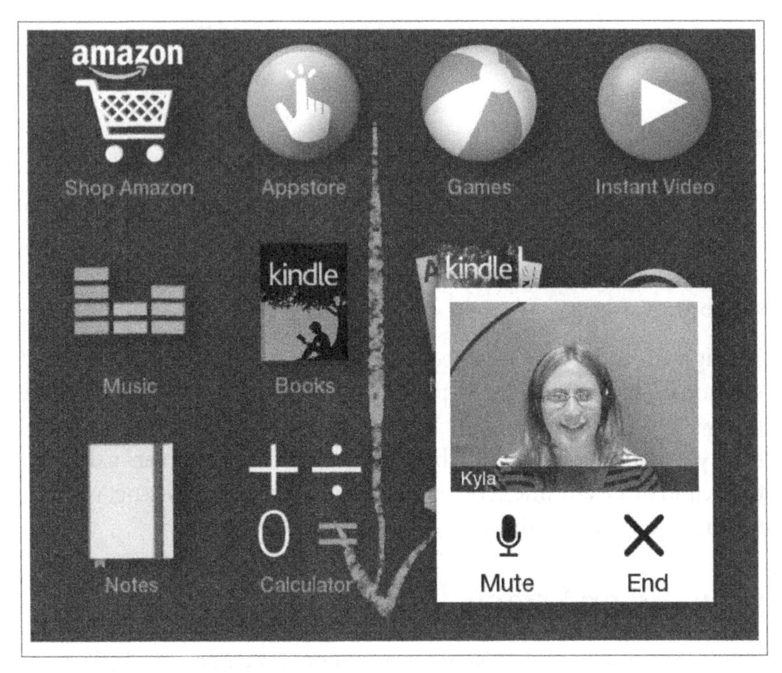

Figure 4-3. An Amazon Tech Advisor pointing out the location of the Fire's Home button

Sometimes, even though you and the Tech Advisor are looking at the same screen, communication about what you're looking at has the potential to break down or become confusing. Perhaps you're viewing a picture in the Photos app, and the Tech Advisor tells you to "tap the Share icon at the bottom of the screen." If you don't recognize that icon that looks like two line segments connected at one point in a sideways V shape (and if you aren't "peeking" at the word "Share" that appears under the icon only when you tilt the phone), you might not know what to press. In this case, the Tech Advisor can simply circle the icon on your screen, and you'll know exactly what she's talking about.

Remote Control

If, try as she might to help talk you through the problem, and if drawing on your screen doesn't help you perform the desired action on your own, Tech Advisors can actually step in to do it for you. Yes, not only can the Tech Advisor *see* your screen and *draw* on it, but she can also operate your device via remote control.

Say you want to assign unique ringtones to your contacts, but you just can't seem to hunt and peck through the Settings→Sounds & Notifications→"Select ringtones for specific people" options, as shown in Figure 4-4. Acting as your online copilot, a Tech Advisor can take over the contols and dig through these preferences for you.

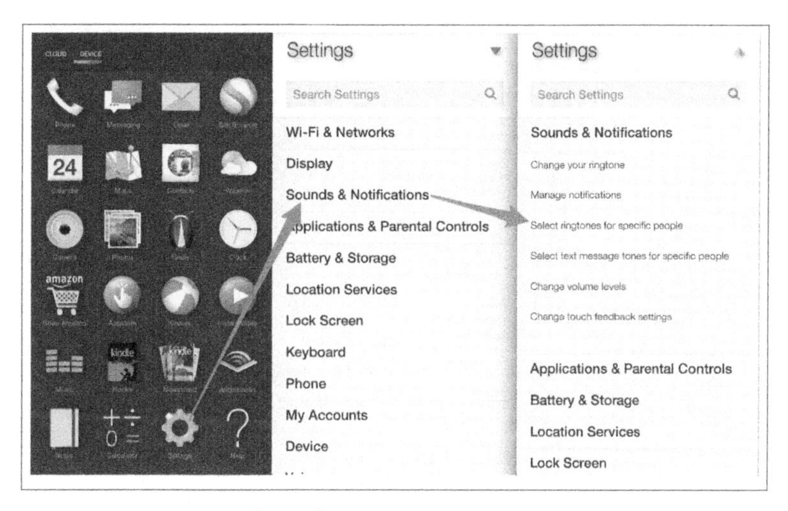

Figure 4-4. Digging through Settings options to assign unique ringtones to contacts

While having someone else do the difficult task for you is helpful on its own, being able to *watch* her do it also teaches you the information for future reference. Now that you've seen her assign ringtones to your contacts' incoming calls, perhaps you won't need to go to Mayday when you later want to select tones for text messages for other contacts.

CHAPTER 5
Camera and Photos

For more and more people, a smartphone is increasingly becoming the primary camera used for most photography, which makes sense, since it's always on you and available when you need it. So, it's important that the camera on the phone is good, and phone specs are becoming significant features in most marketing for new devices.

Fortunately, though knowing the specs doesn't always tell you all that much (large megapixel count doesn't always translate into high-quality photos), Fire's phone is high quality, competitive with specs, and, most importantly, has unique features that really make it stand out. Again, since this guide assumes you've used a smartphone camera before, this chapter will focus on those new and innovative features, combined with unique benefits of the associated Photos app and related Amazon services.

Camera Specs

For the record, the Fire phone has a 13MP rear-facing camera, LED flash, f/2.0 lens, 2.1MP front-facing camera, and the rear-facing camera records video in 1080p HD.

But first, a quick pointer to how to take a picture quickly. Just presss the Camera button (see Figure 1-1) quickly (if you hold it down too long, you'll open the Firefly app) to bring up the camera from anywhere (within an app, from the lock screen, or while the screen is asleep). Press it again (or tap the shutter icon, shown in Figure 5-1 to take a picture.

Figure 5-1. The shutter icon in the Camera app

Now that you can take a picture, on to the new features.

Optical Image Stabilization

Though it's not something you need to do anything in order to use it, it's good to know that the camera makes use of optical image stabilization (OIS) under the hood.

OIS works by means of tiny electric motors on the lens that stabilize it by adjusting 100 times per second. These responsive movements counteract any innevitable shaking of your hand while you're holding the phone, allowing the camera to keep the shutter open up to four times longer to gather more light and information. The result is a clearer, crisper picture with more accurate colors and detail.

Unlimited Cloud Storage

As a benefit to Fire phone owners, Amazon provides completely free, unlimited online photo storage and backup via Amazon Cloud Drive. Enabled by default, the Fire phone automatically sends every picture you take to Cloud Drive.

This means tons of storage for all of your photos, but it also means never having to worry about accidentally deleting a precious memory from your phone and losing it forever. Even if your phone is destroyed after taking the picture, all you need to do to recover it is to go to the

Photos & Videos section of your Amazon Cloud Drive (*https://www.amazon.com/gp/photos*), as shown in Figure 5-2.

Figure 5-2. Photos in Amazon Cloud Drive

From here, you can just select a photo and click the Download button at the bottom-left side of the screen to store the photo on your desktop or other device.

See the picture of a cat in the lower-right corner of Figure 5-2? Though the triangle icon on it might suggest a video, this is actually a *lenticular photo*, another new, exclusive feature on the Fire phone, discussed in "Lenticular Capture Mode" on page 42.

As you can see in Figure 5-2, Fire sends more than just your precious memories to Amazon Cloud Drive. It also sends your terrible mistakes. If you see a photo you know you'll never want to see again, just select it and click the Delete button, which is right next to Download in the bottom-left side of the photo preview page.

Of course, you don't need to go through the trouble of visiting your Cloud Drive just to find and delete a photo. If you see a photo on your phone you know you want to remove forever, just select it within the Photos app and tap the delete icon, as shown in Figure 5-3.

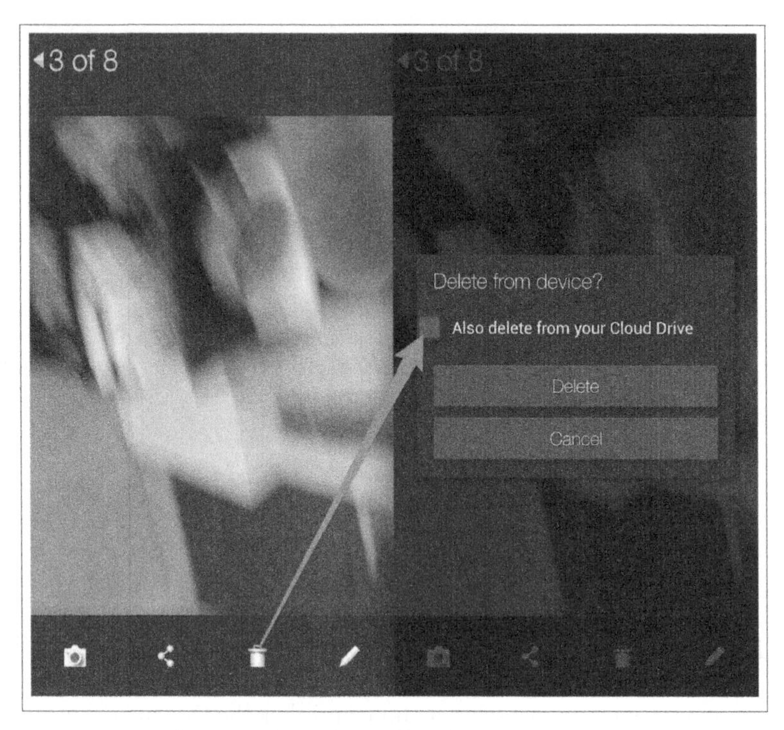

Figure 5-3. Deleting a photo from Cloud Drive

On the screen that opens, just enable the "Also delete from your Cloud Drive" box (as shown in Figure 5-3), tap Delete, and it will be removed from both your phone and Cloud Drive.

Lenticular Capture Mode

After launching the Camera app and tapping the Settings icon, you'll see a variety of options, as shown in Figure 5-4. While most of these (such as HDR, Image Review, and Panorama) have become pretty standard fare on most feature-rich smartphones (so we won't cover them here), the Lenticular capture mode is new and exclusive to the Fire phone.

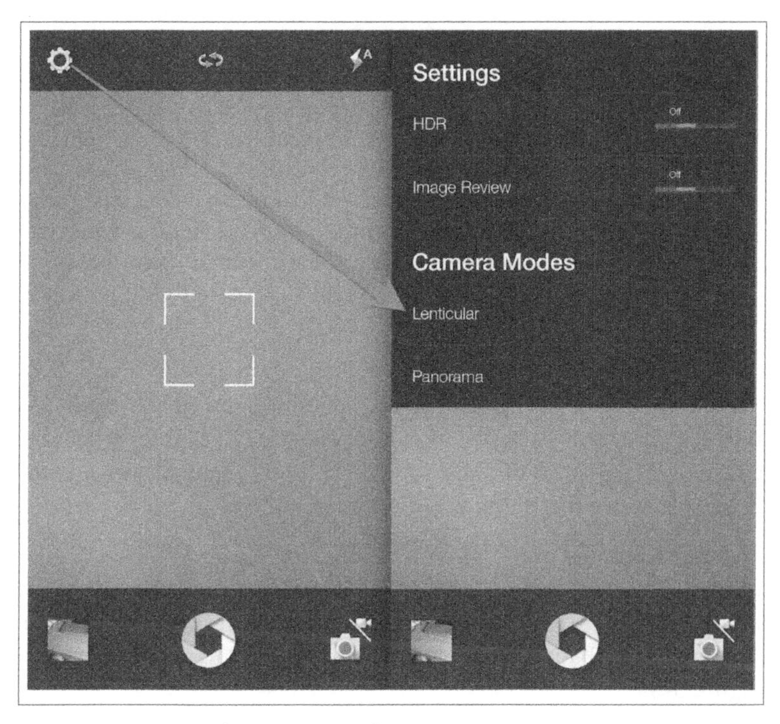

Figure 5-4. Lenticular setting in the Camera app

While it's an interesting feature in itself, if you've read about how Amazon describes the Fire's Dynamic Perspective lock screens, the use of the term *lenticular* (the adjective Amazon uses to describe the 3D look caused by Dynamic Perspective in the lock screens that ship with the phone) can seem a little misleading and disappointing. That is, you should not expect to be able to use this feature to take a spectacular image that makes use of Dynamic Perspective to add to your lock screen. In fact, you won't be able to add it to your lock screen as a shifting-perspective image at all.

If you select your lenticular image within Lock Screen settings, it will install on the lock screen as a single image only (the first image shot in the sequence). Like everything else in the Fire interface, it will shift under the layers of crhome on top of it (in this case, the current date and time, which cast a shadow on the image beneath them), but it will not transition between different perspectives on the image as you might expect. Photos shot in Lenticular mode are also limited to square format, which makes them less than ideal for a lock screen image.

All that said, once you've gotten over the disappointment of what this mode *is not*, it's still a pretty fun and interesting option for what *it is*: a way to take a series of photos and stich them together into an animated sequence that shifts as you tilt the phone.

To take a lenticular photo, open the Camera app and tap the Lenticular camera mode, as shown in Figure 5-4. You'll be prompted to take the first photo in the series (using either the on-screen shutter icon or the Camera button), followed by any others you want to add, as shown in Figure 5-5.

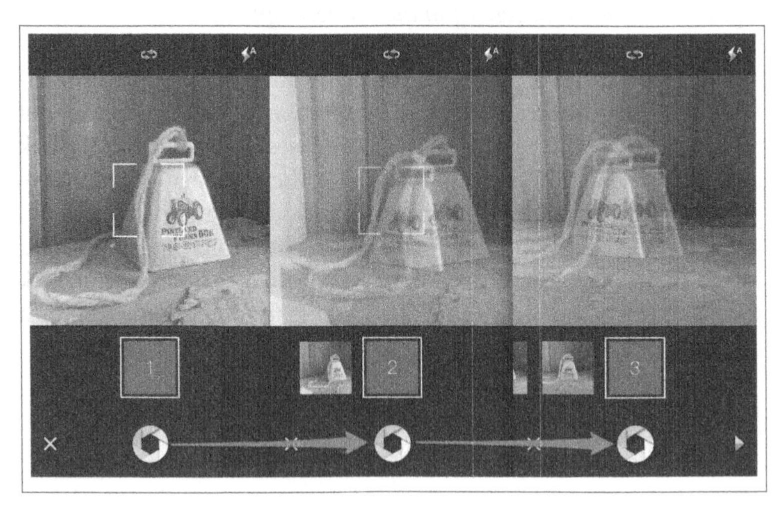

Figure 5-5. Taking a lenticular photo

When you've finished taking as many images as you want to stitch together, tap the triangle at the bottom-right side of the camera screen (shown in Figure 5-5) to save the file (tapping the X in the bottom-right corner deletes all images stored the temporary file completely) to your Photos library.

The result looks like Figure 5-6, shifting from one image in the series to a neighboring image as you tilt your phone.

Figure 5-6. A lenticular photo

If you'd like share your lenticular image with other apps or people, it will be formatted as an auto-playing animated GIF in any web browser.

About the Author

Brian Sawyer is a Senior Editor at O'Reilly Media, where he manages the Missing Manuals division. He is also the author of *Kindle Fire: Out of the Box* and coauthor of *NOOK Tablet: Out of the Box* and *Best Android Apps*. You can him find him on Twitter at @briansawyer.

Colophon

The cover fonts are Benton Sans, Guardian Sans, and Helvetica. The text font is Adobe Minion Pro; the heading font is Adobe Myriad Condensed; and the code font is Dalton Maag's Ubuntu Mono.

Lightning Source UK Ltd.
Milton Keynes UK
UKOW06f1946280515

252458UK00015B/137/P